Ruby-Throated Hummingbird Migration

By Susan H. Gray

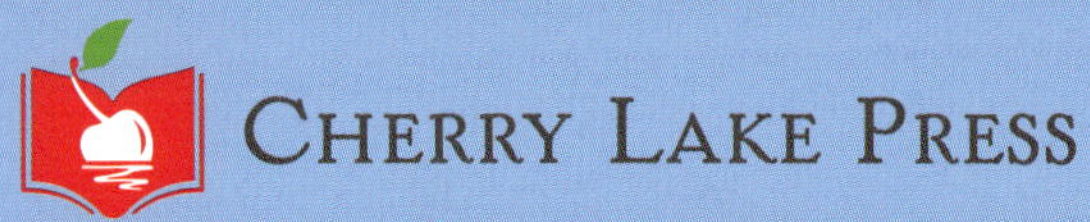

Published in the United States of America by Cherry Lake Publishing Group
Ann Arbor, Michigan
www.cherrylakepublishing.com

Reading Adviser: Marla Conn, MS, Ed., Literacy specialist, Read-Ability, Inc.
Content Adviser: Sheila K. Schueller, Ph.D., Lecturer and Academic Program Specialist, University of Michigan School for Environment & Sustainability
Photo Credits: ©KellyNelson/Shutterstock.com, cover, 4; ©Bonnie Taylor Barry/Shutterstock.com, 6; ©Steven Russell Smith Ohio/Shutterstock.com, 8; ©Fiona M. Donnelly/Shutterstock.com, 10; ©Agnieszka Bacal/Shutterstock.com, 12; ©Dmitry Demkin/Shutterstock.com, 14; ©Daniel Lamborn/Shutterstock.com, 16; ©Ramona Edwards/Shutterstock.com, 18; ©Sari ONeal/Shutterstock.com, 20

Cherry Lake Press is an imprint of Cherry Lake Publishing Group.

Library of Congress Cataloging-in-Publication Data

Names: Gray, Susan Heinrichs, author.
Title: Ruby-throated hummingbird migration / by Susan H. Gray.
Description: Ann Arbor, Michigan: Cherry Lake Publishing, [2021] | Series: Marvelous migrations | Includes index. | Audience: Grades 2-3
Identifiers: LCCN 2020002789 (print) | LCCN 2020002790 (ebook) | ISBN 9781534168565 (hardcover) | ISBN 9781534170247 (paperback) | ISBN 9781534172081 (pdf) | ISBN 9781534173927 (ebook)
Subjects: LCSH: Ruby-throated hummingbird–Migration–Juvenile literature.
Classification: LCC QL696.A558 G727 2021 (print) | LCC QL696.A558 (ebook) | DDC 598.7/641568–dc23
LC record available at https://lccn.loc.gov/2020002789
LC ebook record available at https://lccn.loc.gov/2020002790

Cherry Lake Publishing Group would like to acknowledge the work of the Partnership for 21st Century Learning, a Network of Battelle for Kids. Please visit http://www.battelleforkids.org/networks/p21 for more information.

Printed in the United States of America
Corporate Graphics

CONTENTS

A hummingbird's heart beats
more than 1,000 times a minute when it flies.

A Great Summer

What a summer it has been for the ruby-throated hummingbirds! They arrived at their **breeding grounds** in May. Males showed off their fancy flying skills to females, hoping to attract mates. Birds paired up. Females crafted **miniature** cup-like nests. Teensy, bean-sized eggs were laid. Babies hatched, grew up, and learned to fly. Things had been great here in the United States

Only males have a ruby throat.

and Central Canada. But it couldn't last forever. Summer is ending soon, and it's time to move on.

In August, the tiny birds begin an amazing **migration**. They leave their summer home and head to Mexico and Central America. Many other birds also migrate south around this time. But hummingbirds are unique. They burn energy really fast!

Ask Questions!

A ruby-throat's breast muscles account for at least one-fourth of its body weight. Why is this?

When migrating, a hummingbird has to eat every 10 to 20 minutes.

Fattening Up

Hummingbirds do what any good athlete would do—they prepare their bodies. Normally, a ruby-throat weighs 0.1 to 0.12 ounces (2.8 to 3.4 grams), barely more than a penny. It eats often during the day. For some of its meals, the bird visits flowers. It prefers red and orange trumpet-shaped blossoms. It pokes its beak deep into the flower, shoots out its tongue, and

While sipping nectar, hummingbirds help pollinate many flowers.

laps up the **nectar** inside. Nectar gives the hummingbird quick energy.

Nectar is not their only food. They also eat tree sap, insects, and spiders. Insects and spiders contain fat. As migration time nears, the adult birds double their body fat. With all this extra stored energy, hummingbirds are ready for their trip.

Make a Guess!

Hummingbirds get sap from holes drilled into trees by woodpeckers. Why do woodpeckers drill these holes?

The females feed and take care of the chicks.

Time to Go

When the days grow short and the air cools, it's time to head south. Hummingbirds do not travel in flocks as some birds do. Instead, they usually fly alone. As they

Think!

Why do hummingbirds store up fat before their trip? How might it help? How would it make things more difficult?

Big birds like hawks can be predators of hummingbirds.

travel, the birds catch and eat insects in midair. They also take breaks to feed in gardens or fields of wildflowers. From time to time, they grab a few days to rest and **replenish**.

Every animal migration involves danger. **Predators** could be lurking. Food might be in short supply. Storms could drive migrants off course. Hummingbirds face all of these threats. But the biggest challenge for ruby-throats is the Gulf of Mexico.

By the time they reach the Gulf, they may have flown more than 1,000 miles

Hummingbird eggs are the size of jellybeans!

(1,609 kilometers) already. But it's another 500 miles (805 km) over the Gulf and into Mexico. No gardens, no wildflowers, no tree sap. How can they make it?

Some birds avoid the Gulf by taking a southbound path over Texas. But most birds brave the trip over the water. They build up their fat before heading out. They snag insects as they fly. They stop and rest on fishing boats and oil rigs. Sometimes, oil rig workers set up feeders on the rigs! Finally, the little birds reach land.

Ruby-throated hummingbirds begin their molt in the summer.

Winter Break

During the winter months, much of North America sees cold temperatures, ice, and snow. But in Mexico and Central America, the ruby-throats stay warm and comfortable.

Look!

Search for **molting** hummingbirds on the internet. These birds will look ragged because they are shedding old feathers and growing new ones. Why is it important for birds to molt?

You can set up feeders around your home! Make sure to be responsible and follow directions carefully.

They also make the most of a plentiful food supply. Around February, they start packing on the fat again. It will soon be time to head north.

In March, April, and May, ruby-throats are on the move. They fly about 20 miles (32 km) a day and make it to their summer homes by June. They settle again in the United States and Central Canada. And the cycle repeats. Males pair off with females. Nests and eggs appear. New little birds hatch and learn to fly. It won't be long before August arrives and another migration will begin.

GLOSSARY

breeding grounds (BREED-ing GROUNDZ) areas where animals go to mate

migration (mye-GRAY-shuhn) movement of animals from one region to another and back again

miniature (MIN-ee-uh-chur) very small

molting (MOHLT-ing) shedding old feathers and growing new ones

nectar (NEK-tur) sugary liquid produced by flowers

predators (PRED-uh-turz) animals that hunt and eat other animals

replenish (ree-PLEN-ish) to fill up again

FIND OUT MORE

BOOKS

Arnold, Quinn M. *Hummingbirds*. Mankato, MN: Creative Paperbacks, 2017.

Burleigh, Robert. *Tiny Bird: A Hummingbird's Amazing Journey*. New York, NY: Henry Holt and Co., 2020.

Sill, Cathryn. *About Hummingbirds: A Guide for Children*. Atlanta, GA: Peachtree Publishing Co., 2015.

WEBSITES

BioKids Critter Catalog—Ruby-Throated Hummingbird
http://www.biokids.umich.edu/critters/Archilochus_colubris
Read about the behavior, life span, and courtship of ruby-throated hummingbirds—and more.

World of Hummingbirds—Hummingbird Facts
http://www.worldofhummingbirds.com/facts.php
Scroll through a huge list of fun facts about hummingbirds and click on the links to learn more.

INDEX

ABOUT THE AUTHOR

Susan H. Gray has a master's degree in zoology. She has written more than 170 reference books for children and especially loves writing about animals. Susan lives in Cabot, Arkansas, with her husband, Michael, and many pets.